Florida's Burmese Pythons

Squeezing the Everglades

by Miriam Aronin

Consultant: Michael E. Dorcas, PhD
Professor of Biology
Davidson College
Davidson, North Carolina

BEARPORT
PUBLISHING

New York, New York

Credits

Cover and Title Page, © jenjen42/iStock and © John A. Anderson/Shutterstock; 4T, © mlorenz/Shutterstock; 4B, © Steffen Foerster/Shutterstock; 5L, © MP cz/Shutterstock; 5RT, © Thomas Luhring; 5RB, © Frank Mazzotti; 6L, © The Design Lab; 6R, © Irina oxilixo Danilova/Shutterstock; 7, © AP Photo/Pat Carter; 8, tinyurl.com/n4v288a/CC-BY-SA-2.0; 9, © Paul Tessier/Shutterstock; 10T, © Allen McGregor/tinyurl.com/phpydbq/CC-BY-2.0; 10B, © Anthony Ricci/Shutterstock; 11, © Heiko Kiera/Shutterstock; 12, Everglades NPS/R. Cammauf/tinyurl.com/olqn57c/CC-BY-2.0; 13, © John Mitchell/Science Source; 14, © iStockphoto/FloridaGator; 15, Lori Oberhofer/National Park Service; 16, © ZUMA Press, Inc/Alamy; 17, Lori Oberhofer/National Park Service; 18, Lori Oberhofer/National Park Service; 19L, © Patrick K. Campbell/Shutterstock; 19R, Catherine Puckett/USGS; 20, © Florida Museum photo by Kristen Grace; 21, © Florida Museum photo by Kristen Grace; 22, © XNR Productions, Inc.; 23T, © MP cz/Shutterstock; 23B, © ZUMA Press, Inc/Alamy; 24, © ZUMA Press, Inc/Alamy; 25, © ZUMA Press, Inc/Alamy; 26, Clay DeGayner/USFWS/tinyurl.com/mm24rre/CC-BY-2.0; 27, © ZUMA Press, Inc/Alamy; 28T, © MartinMaritz/Shutterstock; 28B, © Ivan Kuzmin/Shutterstock; 29T, © Arnoud Quanjer/Shutterstock; 29B, © Audrey Snider-Bell/Shutterstock.

Publisher: Kenn Goin
Editor: Jessica Rudolph
Creative Director: Spencer Brinker
Design: The Design Lab
Photo Researcher: Jennifer Zeiger

Library of Congress Cataloging-in-Publication Data

Aronin, Miriam, author.
Florida's Burmese pythons : squeezing the everglades / by Miriam Aronin.
pages cm. — (They don't belong : tracking invasive species)
Summary: "In this book, readers learn how the Burmese Python is an invasive species in the state of Florida."—Provided by publisher.
Audience: Ages 7–12
Includes bibliographical references and index.
ISBN 978-1-62724-827-3 (library binding) — ISBN 1-62724-827-7 (library binding)
1. Burmese python—Juvenile literature. 2. Pest introduction—Florida—Everglades—Juvenile literature. 3. Nonindigenous pests—Florida—Everglades—Juvenile literature. 4. Nature—Effect of human beings on—Juvenile literature. 5. Everglades (Fla.)—Juvenile literature. I. Title.
QL666.O63A76 2016
597.96'78—dc23
2015009028

For more information, write to Bearport Publishing Company, Inc., 45 West 21st Street, Suite 3B, New York, New York 10010. Printed in the United States of America.

10 9 8 7 6 5 4 3 2

Contents

A Mammal Mystery

In 2012, a team of scientists announced shocking news. They had discovered that since the 1990s, there were far fewer **mammals** in southern Florida. The scientists found very few bobcats in the area. Raccoons and opossums had almost disappeared. They found no rabbits or foxes at all. What could be causing these animals to vanish?

A bobcat

A raccoon

The scientists had been studying mammals living in a huge **national park** in Florida called the Everglades. In 2012, scientists found 87 percent fewer bobcats and 99 percent fewer raccoons in the area than in the 1990s.

The scientists had an idea—a large snake called a Burmese python might be to blame. After all, the giant snakes can **devour** almost any mammal they come across. One of the scientists, Dr. Frank Mazzotti, described pythons as "vacuum cleaners on mammals."

A Burmese python

Dr. Michael Dorcas led the team of scientists that studied wildlife in southern Florida.

Dr. Frank Mazzotti was part of Dorcas's team.

Unusual Pets

What's really surprising is that Burmese pythons are not **native** to Florida. So how did they get there? The snakes originally come from southern Asia. For several decades, people have been shipping young Burmese pythons from Asia to the United States to be sold as pets.

Where Burmese Pythons Come From

Where Burmese pythons live in Asia

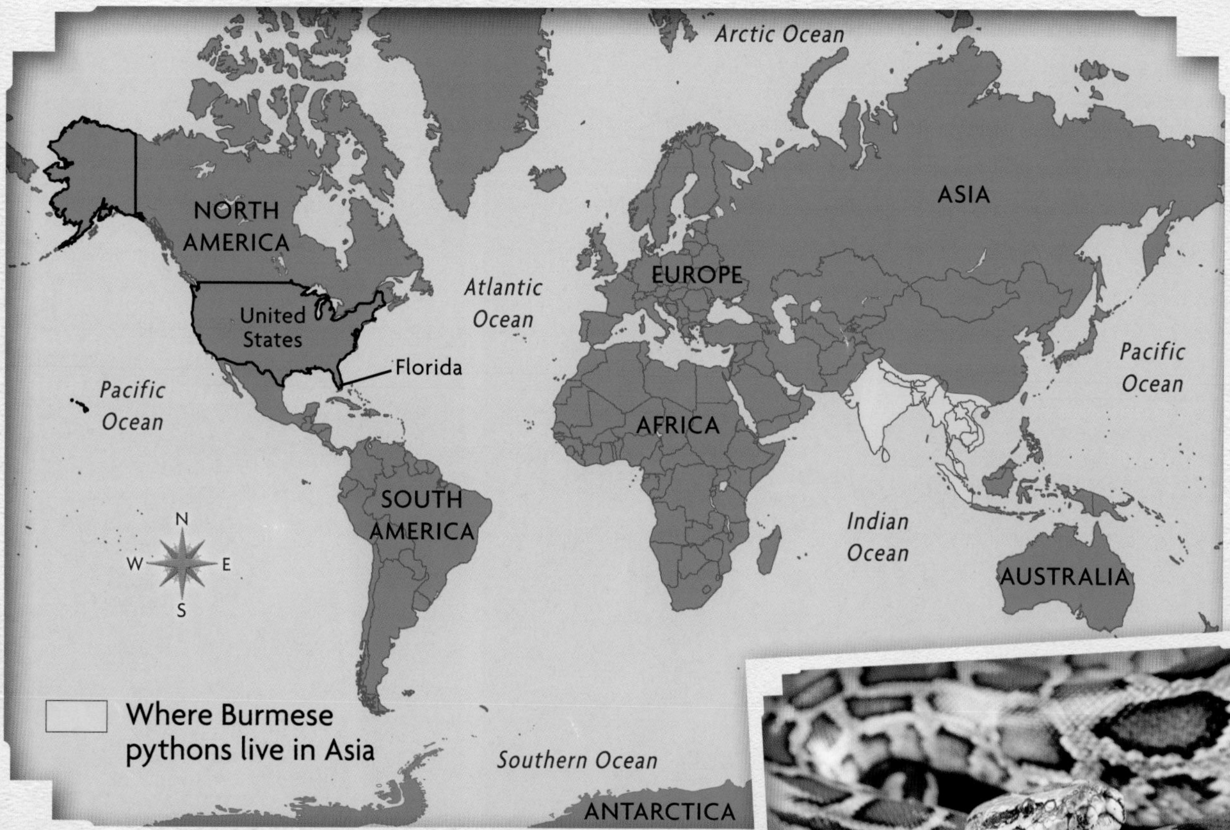

Some areas of southern Asia that Burmese pythons come from are warm and rainy. Other areas are dry and get cold during winter.

As babies, the snakes are around 2 feet (61 cm) long, and they're usually gentle. Pet owners house their pythons in **terrariums** and feed them mice. However, the cute baby snakes grow quickly and eventually become giants. Soon, the **reptiles** need a lot more food. They also become much too large for regular terrariums. Instead, each adult snake needs a space as big as a small room.

Young pythons, like the one shown above, are usually gentle, but there is a risk of adult pythons attacking their owners.

Burmese pythons reach their full size when they are four years old. Adults can grow to a length of more than 20 feet (6.1 m).

Into the Everglades

In the mid-1990s, scientists made a startling discovery. They found Burmese pythons living in Everglades National Park, a huge protected area for plants and animals. How did the pythons end up in the wild? No one knows for sure. However, scientists have learned that many snake owners find caring for their growing pythons difficult and expensive. A few of these owners may have let their giant pets go free. It's also possible that some pythons escaped on their own.

More than one million people visit Everglades National Park every year. Visitors enjoy activities such as hiking, biking, kayaking, camping, and bird watching.

One thing was certain. Burmese pythons had found a comfortable home in the Everglades. Even worse, the **population** of this **invasive species** was growing. In 2006, scientists discovered a nest full of python eggs in the park. That meant the snakes were **breeding**. Soon, more and more pythons would be roaming the Everglades!

After laying eggs in spring, a female python stays coiled around the eggs to keep them warm. After about two months, the eggs hatch, and the young snakes leave the nest.

A Perfect Home

For Burmese pythons, Everglades National Park is a perfect place to live. The park gets lots of rain and is warm all year. In fact, the **climate** in the Everglades is very similar to the climate in many parts of the snakes' native home in Asia.

The Everglades is home to more than 300 kinds of birds, including this anhinga.

Everglades National Park covers more than 1.5 million acres (607,028 ha) in southern Florida. The average temperature there is around 82°F (28°C). The park

The Everglades also has many kinds of **habitats** where the snakes can easily live, such as grasslands, forests, and wetlands. Pythons slither along the ground, climb trees, and swim in lakes and ponds. This allows them to find **prey** throughout the park.

A Powerful Hunter

As it hunts, a Burmese python's skin blends in with plants and murky water. This **camouflage** allows the snake to stay hidden as it looks for an animal to eat—either by following its prey or waiting for it to get close. When an animal is nearby the snake attacks! The python's **victim** may not see or hear the snake until it's too late.

A python in a tree

Some snakes kill their prey with **venom**, but Burmese pythons do not. Instead, a python uses its teeth to grab the prey before wrapping its body around the animal. Then, the powerful reptile constricts, or squeezes, the prey very hard. This causes the animal to stop breathing, and soon it dies.

After killing it's prey, a Burmese python swallows the animal whole.

A Burmese python killing its prey

Top Predators

Burmese pythons kill all kinds of creatures—large and small. The snakes often hunt small birds, rabbits, and rats. However, they can also take down large animals such as deer. This can be a problem for other **predators**, such as bobcats and Florida panthers. Because pythons eat many of the same kinds of prey that these animals eat, there may be less food available for the native predators.

A deer

Burmese pythons have spread to other areas of southern Florida, outside of the Everglades. This means even more native animals in the state are in danger.

Another reason pythons are so destructive is that they are a top predator. This means there are few animals that hunt adult pythons—and, as a result, control their population. In Florida, only large alligators have been able to kill fully grown pythons. Yet, sometimes, when pythons cross paths with alligators, the snakes still win!

An alligator struggling to kill a Burmese python

Gut Check

Scientists want to find ways to fix the Burmese python problem in Florida. First, however, they have to learn more about the snakes. To do so, scientists study the bodies of pythons that have been captured and killed.

A scientist holds a python's head tightly so the snake cannot bite.

Sometimes scientists get help catching pythons from expert python hunters. Hunters and scientists use their bare hands and long hooks to catch the snakes. Then they put the captured pythons into special bags or boxes.

To learn about the different kinds of prey that pythons eat, for example, scientists can look inside the snakes' stomachs. Experts have discovered more than 40 kinds of mammals and birds inside pythons. Some of these animals are **endangered**. If the pythons are not controlled, these animals could become **extinct**.

This large bird was found inside a Burmese python's stomach.

Tracking Monsters

Scientists also study how pythons behave in the wild. They **implant** devices, such as **radio transmitters** and **motion sensors**, into the bodies of captured snakes. Then they release the pythons back into the wild.

Radio transmitters track exactly where the snakes go. Motion sensors help scientists figure out the size of the pythons' prey. For example, a python can squeeze the breath out of small prey very quickly. However, a snake takes more time and moves more when it wrestles and kills a large animal such as an alligator.

Researchers implant a radio transmitter into a Burmese python.

On March 6, 2012, a team of scientists captured a huge female python in the Everglades. It was 17 feet 7 inches long (5.4 m) and weighed 164.5 pounds (75 kg). The scientists implanted a motion sensor and two radio transmitters into the monster snake. Then they released her into the wild and waited a few weeks before recapturing her.

An eastern indigo snake

The largest native snake in Florida is the eastern indigo snake. It can grow up to 8 feet (2.4 m) long. That's less than half the length of the huge Burmese python that scientists caught in March 2012.

A radio transmitter implanted into the side of a Burmese python

Inside the Giant

On April 19, 2012, with the help of the radio transmitters, the scientists found and caught the same large female snake. This time, they did not let her go. Instead, they killed her and cut open her body. They couldn't believe what they found inside the snake—87 eggs she was about to lay! That was the most eggs ever found inside a Burmese python in the Everglades.

Scientists cut open the body of the giant snake they recaptured.

Luckily, the scientists caught the snake before she could lay her eggs. Still, the python was alarming. Her huge size showed that she was getting plenty to eat in the Everglades. Scientists realized the python problem was bigger than they had thought. Reptile expert Dr. Kenneth Krysko said, "These snakes are surviving a long time in the wild. There's nothing stopping them, and the native wildlife are in trouble."

A scientist removing eggs from the snake's body

egg

After studying Burmese pythons for many years, scientists still aren't sure how many of them live in Florida. That's because the snakes are extremely good at staying hidden. There may be a few thousand or hundreds of thousands of pythons in the state.

Python Patrol

Today, people are working on ways to control the Burmese python problem, including passing new laws. It is now illegal to bring Burmese pythons into the United States. It is also against the law to buy a pet python in Florida.

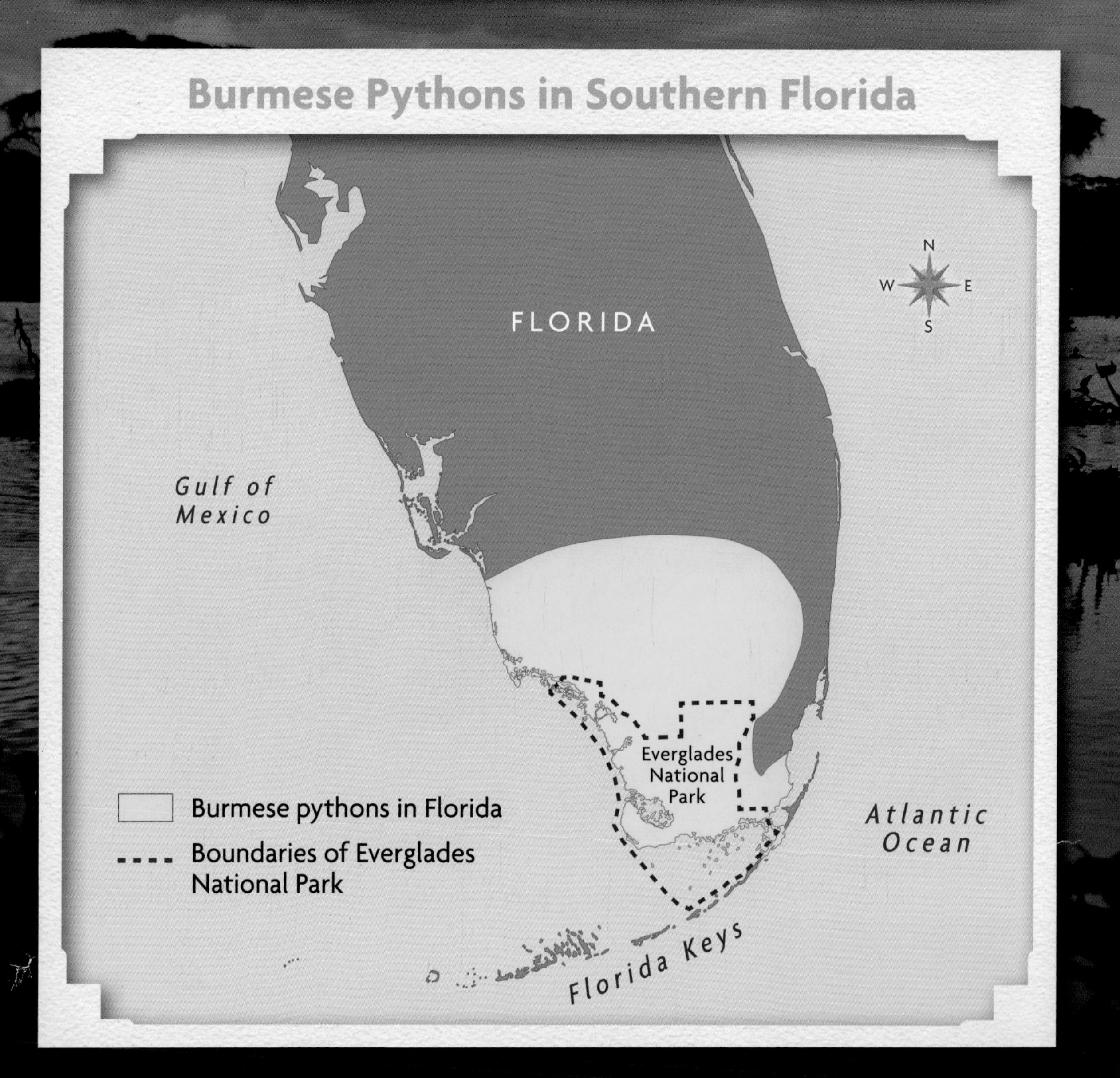

Most Burmese pythons living in the wild in the United States are in southern Florida.

Volunteers that work for a Florida program called Python Patrol also help. Patrol members first learn to identify Burmese pythons. Then, if they spot a python in the wild, they report it. After that, snake experts or volunteers with more training capture the python. By 2014, more than 3,000 people had joined the Python Patrol.

Python Patrol volunteers are taught to look at the head of snakes to tell which ones are Burmese pythons. A python's head is shaped like a pyramid. On top, there is a dark mark shaped like an arrowhead.

Some volunteers learn to safely capture Burmese pythons by grabbing them by the tail and behind the head. Then the volunteers give the snakes to experts who have special permission to kill and study them.

The Python Challenge

In 2013, officials in Florida found another way to help control the python population. The state held a contest called the Florida Python Challenge. Whoever caught the most pythons in one month would win $1,500. Around 1,600 people took part in the contest, including a man named Ruben Ramirez.

During the Python Challenge, anyone who signed up and took a training class was allowed to hunt the big snakes.

One day during the contest, Ruben spotted a 9-foot (2.7 m) python next to a **canal**. The huge snake tried to escape to the nearby woods. It was fast, but Ramirez was faster, and he caught the snake by the tail.

By the end of the contest, a total of 68 snakes had been caught. Ruben alone had caught 18 pythons—more than any other person. That meant he won the contest.

Ruben Ramirez (standing) watches as officials measure a Burmese python he caught during the Python Challenge.

The Python Challenge didn't reduce the number of pythons in Florida by much. However, news reports about the contest taught many people not to release Burmese pythons and other invasive species into the wild.

Looking Ahead

What's the future of Burmese pythons in Florida? Scientists know it may be impossible to get rid of all the pythons, so their present goal is to keep pythons from spreading to more places in the state. Unfortunately, scientists have recently found several pythons in the Florida Keys. These are islands near the state's southern coast. One captured snake had two endangered Key Largo woodrats in its stomach.

A Key Largo woodrat

Still, there is hope that ordinary Floridians can help limit the spread of Burmese pythons to new areas. The Florida government has a special phone number, a website, and a smart phone app that people can use to let officials know where they have spotted the snakes. By reporting pythons they see, people can help protect Florida's native wildlife.

This Burmese python was captured in Florida in 2013. While most pythons have dark-colored patches on their skin, some, like this snake, have yellow patches.

No visitors to the Everglades have ever been attacked by a Burmese python. However, pythons can be dangerous, so it's best to stay away from them.

Other Invasive Reptiles

Florida has more invasive species than any other state. Many of the animals were pets that escaped or were released into the wild. Here are four more invasive reptiles found in Florida.

African Rock Python

- African rock pythons come from Africa. They have been found in areas around Miami, Florida.
- The pythons can grow to 20 feet (6.1 m) in length. They are the largest snakes in Africa.
- African rock pythons squeeze their prey to death. They eat many kinds of animals, including several endangered species in Florida.
- If they are bothered, the snakes may attack people with their long, sharp teeth.

Boa Constrictor

- Boa constrictors come from Central America and South America. They have been found in a park in Miami, Florida.
- Boa constrictors are good climbers. They can live on land or in trees.
- Boas can grow up to 13 feet (4 m) long.
- Females do not lay eggs. Their young are born live.
- The snakes kill their prey by squeezing. They eat reptiles, birds, and small mammals.

Nile Monitor Lizard

- Nile monitor lizards come from Africa. They have been found in several coastal areas of Florida.
- A Nile monitor may grow to more than 7 feet (2.1 m) long. It's the longest lizard in Africa.
- The lizards can run up to 18 miles (29 km) per hour. They are good swimmers, too. They can stay underwater for up to an hour.
- Nile monitors eat small mammals, reptiles, frogs, and birds. They also eat bird, alligator, and turtle eggs.
- Nile monitors often live in the same areas as people. Some lizards have gotten into people's homes, and even killed and eaten people's dogs and cats.

Tegu

- Tegus are lizards that come from South America. Many kinds of tegus have been found throughout Florida, including near Tampa and Miami.
- The lizards can grow up to 5 feet (1.5 m) long.

- Tegus eat plants and small animals. The lizards also eat the eggs of birds and tortoises that build their homes on or in the ground.
- Tegus have sharp teeth and claws. It is not safe for people to capture them by hand. However, they can be caught in traps.

Glossary

breeding (BREED-ing) producing young

camouflage (KAM-uh-flahzh) patterns and colors on an animal's body that help it blend into its surroundings

canal (kuh-NAL) a human-made waterway

climate (KLYE-mit) the typical weather in a place

devour (di-VOUR) destroy; eat hungrily and quickly

endangered (en-DAYN-jurd) in danger of dying out

extinct (ek-STINGKT) when a kind of plant or animal has died out completely

habitats (HAB-uh-*tats*) places in nature where plants and animals live

implant (im-PLANT) put something in a part of a living body

invasive species (in-VAY-siv SPEE-sheez) plants or animals that have been moved from their habitat into another habitat in which they do not naturally belong

mammals (MAM-uhlz) warm-blooded animals that have a backbone and hair or fur on their skin; they also nurse their babies

motion sensors (MOH-shuhn SEN-surz) devices that can sense the movement of people or animals

national park (NASH-uh-nuhl PARK) an area of land set aside by the government to protect the animals and plants that live there

native (NAY-tiv) naturally born and living in a particular place

population (*pop*-yuh-LAY-shuhn) the number of people or animals living in a place

predators (PRED-uh-turz) animals that hunt and kill other animals for food

prey (PRAY) animals that are hunted and eaten by other animals

radio transmitters (RAY-dee-oh tranz-MIT-urz) objects that send out radio signals and are put on or in an animal so that its movements can be tracked

reptiles (REP-tyelz) cold-blooded animals, such as lizards, snakes, turtles, and alligators, that use lungs to breathe and usually have dry, scaly skin

terrariums (tuh-REHR-ee-uhmz) glass or plastic boxes used for growing plants or keeping small animals indoors

venom (VEN-uhm) poison that some animals, such as snakes, can send into the bodies of other animals through a bite or sting

victim (VIK-tuhm) an animal that is attacked or killed by another animal

volunteers (*vol*-uhn-TIHRZ) people who help others for free

Bibliography

Dell'Amore, Christine. "Biggest Burmese Python Found in Florida—17.7 Feet, 87 Eggs." *National Geographic News* (August 15, 2012).

Dorcas, Michael E., and John D. Willson. *Invasive Pythons in the United States: Ecology of an Introduced Predator.* Athens, GA: University of Georgia Press (2011).

Reimers, Frederick. "Python Patrol." *Nature Conservancy Magazine* (2013).

Read More

Collard, Sneed B., III. *Science Warriors: The Battle Against Invasive Species (Scientists in the Field).* Boston, MA: Houghton Mifflin (2008).

Oachs, Emily Rose. *Burmese Pythons (Amazing Snakes!).* Minneapolis, MN: Bellwether Media (2014).

Spilsbury, Richard. *Invasive Reptile and Amphibian Species (Invaders from Earth).* New York: PowerKids Press (2015).

Learn More Online

To learn more about Burmese pythons, visit
www.bearportpublishing.com/TheyDontBelong

Index

About the Author

Miriam Aronin is a writer and editor in Chicago. She enjoys reading, dancing, knitting, and avoiding invasive species.